Raphaël Simeon Njock

Chemical mutagenesis and genetic improvement of tropical plants

Raphaël Simeon Njock

Chemical mutagenesis and genetic improvement of tropical plants

Use of sodium azide to create mutants from okra seeds

ScienciaScripts

Imprint
Any brand names and product names mentioned in this book are subject to trademark, brand or patent protection and are trademarks or registered trademarks of their respective holders. The use of brand names, product names, common names, trade names, product descriptions etc. even without a particular marking in this work is in no way to be construed to mean that such names may be regarded as unrestricted in respect of trademark and brand protection legislation and could thus be used by anyone.

Cover image: www.ingimage.com

This book is a translation from the original published under ISBN 978-620-6-72450-6.

Publisher:
Sciencia Scripts
is a trademark of
Dodo Books Indian Ocean Ltd. and OmniScriptum S.R.L publishing group

120 High Road, East Finchley, London, N2 9ED, United Kingdom
Str. Armeneasca 28/1, office 1, Chisinau MD-2012, Republic of Moldova, Europe
Printed at: see last page
ISBN: 978-620-8-20799-1

USE OF SODIUM AZIDE TO CREATE MUTANTS FROM PLANT SEEDS: THE SPECIFIC CASE OF OKRA

Raphaël Simeon NJOCK[1*]

[1] Université de Yaoundé 1, Faculté des Sciences, Département de Biologie et Physiologie Végétales, Unité de Génétique et Amélioration des Plantes, BP 812 Yaoundé, Cameroon

**Corresponding author: simeonnjocky@gmail.com*

SUMMARY

*To date, the induction of mutations from plant seeds by sodium azide (SA) has emerged as a significant alternative for the creation of interesting traits. Because of the satisfactory improvements observed during its use on other plants, and following a weak literature marking its impact in the improvement of the phytochemical response of Abelmoschus esculentus whose nutritional and medicinal importance is no longer to be demonstrated, the present manuscript aims to make a double contribution. It presents an effective assessment of the use of SA, supported by a practical study of trials to improve phytochemical attributes by applying this mutagen to three varieties of okra grown in Cameroon. In a context of perpetual search for solutions to satisfy human food and health needs, the mastery of this technology (**chemical mutagenesis**), newly deployed both regionally (Central Africa) and nationally, comes at just the right time. It represents a veritable golden niche for the rapid creation of useful, strategic variability, serving as a solid basis for the many choices made by breeders.*

CHAPTER I

REVIEW OF THE CREATION OF MUTANTS BY APPLYING SODIUM AZIDE TO PLANT SEEDS

Raphaël Simeon NJOCK[1*] , Benoit Constant LIKENG-LI-NGUE [1,3] , Luther Fort MBO NKOULOU[2] , Hermine Bille NGALLE[1] and Martin Joseph BELL [1]

[1] Université de Yaoundé 1, Faculté des Sciences, Département de Biologie Végétale, Laboratoire de Génétique et Amélioration des Plantes, BP 812 Yaoundé, Cameroon

[2]Institute of Agricultural Research for Development (IRAD), BP 2123 Yaoundé, Cameroon

[3]University of Yaoundé 1, Faculty of Science, Department of Plant Biology, Centre de Recherche et d'Accompagnement des Producteurs Agro- pastoraux du Cameroun (CRAPAC), BP 812 Yaoundé, Cameroon

**Corresponding author: simeonnjocky@gmail.com*

Summary

The aim of this review is to provide an update on the induction of mutations by the application of sodium azide to seeds. Specifically, the methods of application and the appropriation of the various effects induced by this mutagenic agent on plants are highlighted. A comparative documentary analysis of previous work was carried out in order to highlight: its methods of use, the appropriate effective doses of this mutagen, as well as the multiple responses of the plant in morphological, physiological, biochemical, molecular and yield terms. From an investigation of previous studies, it emerged that pre-treating seeds with water for a period of 6 h, combined with exposure to mutagen solution ranges (NaN_3) of: 0; 1; 2; 3; 4; 5 and 6 g/L or 0; 2; 4 and 6 g/L for 2 and/or 4 and/or 6 h, would be ideal for inducing mutations in the genetic material of germinating seeds. In addition, for each parameter assessed and depending on the dose (concentration x exposure time), there was either a reduction, improvement or no change in the effect. These results should make it possible to propose perfectly credible protocols for inducing genetic diversity using this mutagenic agent in species that have not yet been tested.

Key words: sodium azide, mutations, mutagen, pre-treatment

1. Introduction

The drop in crop productivity observed in every part of the world as a result of climate change is coupled with a 60% increase in food demand [1]. This situation is compounded by a sharp rise in the world's population, which is expected to exceed 9 billion by 2050. To meet the world's food needs, agricultural production will have to increase by 70% [2]. This will require a transformation in agricultural systems [1]. To achieve this, crop production will have to adapt to climate change and mitigate its effects [2]. Given the limitations in terms of agricultural performance observed in the varieties already in use, it will be necessary to develop new varieties incorporating a range of interesting characteristics. These varieties will have to be adaptable to new agricultural zones, while offering better resistance to diseases and pests. They should also offer the possibility of growing them outside traditional production periods, while being more efficient in the use of available water. They should also have greater nutritional value and higher yields [3].

These new varieties are obtained by a multitude of methods, including hybridisation, transgenesis [3] and mutagenesis [4]. The particularity of this last method is that it can be carried out on all types of plant material from which an entire plant can be regenerated [5]. This logically increases the possibility o f obtaining mutant progeny (new varieties resulting from induced mutation) with interesting characteristics. A mutation is the primary source of any genetic variation that may exist in several organisms, including plants [6]. A mutation is natural when it occurs as a result of environmental conditions. On the other hand, it becomes induced as soon as the modifications provoked in the of genetic material (genes and chromosomes) are linked to physical and/or chemical mutagenic agents applied by humans [7].

Mutations can affect the genome in a number of ways, including point or widespread modifications. They can also involve variations in the number of chromosomes. According to [8], point mutations are identified by the substitution or transversion of nucleic bases, resulting in missense, nonsense or silent mutations, while extended mutations resulting from crossover phenomena are characterised by changes that can affect a single chromosome (deletion, addition and inversion) or two (translocation). When the induction of numerical changes in chromosomes involves the whole genome, the term euploid variation is used (haploid, diploid, triploid, tetraploid, etc.); on the other hand, when only part of the genome is affected, the term aneuploid variation is used (nullisomic, monosomic, trisomic, tetrasomic, double trisomic) [9].Inducing mutation using physical and/or chemical mutagens is a method that makes it possible to create genetic diversity leading to new varieties with the best characteristics [10]. Traits resulting from induced mutagenesis may be heritable [11]. Physical mutagenesis is characterised by the use of radiation (gamma rays, X-rays, ions, etc.), while chemical mutagenesis uses mainly alkylating compounds such as ethylmethanesulphonate, methylmethanesulphonate, etc., which can be used to induce mutagenesis.[5], and others such as colchicine; the use of sodium azide is not negligible [12].

Chemical mutagens are more effective than physical mutagens in that, as well as increasing genetic variability within plants, they guarantee greater success in selection programmes by vegetative propagation and sexual reproduction [13, 14]. In the latter case, seeds are ideal for inducing mutations. because their life cycle allows mutant generations to be produced rapidly and new lines with desirable traits to be developed [15]. Among the chemical mutagens tested, sodium azide (NaN_3) was found to be one of the most potent used to induce mutation in cultivated plants [16] at a high frequency [17]. The mutagenicity of this agent is mediated by the production of an organic metabolite that penetrates the nucleus, interacts with DNA and creates a point mutation in the plant genome [18]. The general aim of this review is to provide an update on the induction ofmutations by the application of sodium azide (SA) to seeds.

2. How to apply sodium azide to seeds

Mastering the handling of sodium azide in order to improve its effectiveness on plant material inevitably requires knowledge of a number of factors, including: the nature and duration of the pre-treatment, the active concentration of the mutagen, the duration of exposure to sodium azide, the biological specificity of the plant concerned and the correlation between the duration of the treatment and the appropriate concentration of this mutagenic agent.

2-1. Seed pre-treatment

Pre-treatment results in a better response to the mutagen [19, 20]. 38 of the 52 experiments analysed used a pre-treatment (**Table 1**). Of these, water/distilled water was used in the majority (28 out of 52 experiments, i.e. 53.84%). The application time ranged from 4 to 24 hours (4 h at 17.85%; 6 h at 35.71%, 12 at 21.42% and 24 h at 7.14%). At In addition, for each of the pre-treatment substances (water, others), the protocol simply consists of soaking the healthy seeds in a container containing each of them.

2-2. Concentration and duration of exposure of seeds to sodium azide

The mutagen solutions prepared serve as references for the development of a variety of concentration ranges (**Table 1)**. These form the basis for determining the active concentrations associated with the exposure times, so that the appropriate doses (concentration in AI + exposure time) can be determined for the correct use of this mutagen. In this experimental context, various concentration ranges are frequently used, ranging from: 0.001 - 0.004 g/L; 0.065(≈0.07) - 0.26 g/L; 0.1 - 0.3 g/L; 1 - 3 g/L; 2 - 6 g/L (**Table 1**). Based on these ranges, so-called active concentrations that have had an effect on one or more parameters are identified. Among these, the most commonly used in order of increasing value are: 0.002 g/L (at 8.33%); 0.07 g/L (at 11.11%); 0.13 g/L (at 8.33%); 0.3 g/L (at 16.67%); 0.4 g/L (at 19.44 %); 0.5 g/L (at 5.56%); 2 g/L (at 8.33%); 3 g/L (at 11.11%); 5 g/L (at 5.56%) and 6 g/L (at 5.56%). In order to set up experimental doses (concentration x exposure time) that can be used, the above concentrations can be combined with the most commonly used exposure times to sodium azide. These are : 2 h; 4 h and 6 h (**Table 1)**.

Mutations can be induced in scarified and surface-sterilised seeds or in seeds in the active growth phase (germination). Furthermore, the biological material (seeds) involved could come from a multitude of plant families (**Table 3**). However, the mutagenic solutions used to contain the seeds must be freshly prepared at low temperature and never stored [15]. From the heterogeneity of the values presented above, it would be possible to derive doses (concentration x exposure time) that should be used for tests on a multitude of plants. Thus, taking into account all the above ranges and active concentrations, the following groups of concentrations: 0; 1; 2; 3; 4; 5 and 6 g/L then 0; 2; 4 and 6 g/L could be proposed and used in experiments with a view to obtaining a greater effect from this mutagen. Initially, exposure times of 2, 4 or 6 hours could be independently associated with the above concentration groups, as illustrated below:

-0; 1; 2; 3; 4; 5 and 6 g/L for 2 h	or	0; 2; 4 and 6 g/L for 2 h
-0; 1; 2; 3; 4; 5 and 6 g/L for 4 h	or	0; 2; 4 and 6 g/L for 4 h
-0; 1; 2; 3; 4; 5 and 6 g/L for 6 h	or	0; 2; 4 and 6 g/L for 6 h

In addition, these times can be combined with each concentration range for more complex experiments, as shown below:

- 0; 1; 2; 3; 4; 5 and 6 g/L for 2, 4 and then 6 h ;

- 0, 2, 4 and 6 g/L for 2, 4 then 6 h.

Table 1: The table shows the applications of the pre-treatments and doses of sodium azide on the seeds: highlighting the active concentration.

Pre-treatment		Concentrations (C) of sodium azide (g/L) used						C. A	Duration of exposure of seeds to sodium	Plants	Ref.
Nature	Duration	C1	C2	C3	C4	C5	C6				
	6 h	1	2	3	/	/	/	3	1	Abelmoschus esculentus	[21]
	4 h	2	4	6	/	/	/	6	4	Hibiscus cannabinus	[22]
	12 h	1	2	3	4	5	/	5	6	Trigonella foenum-graecum	[23]
	12 h	0,10	0,13	0,16	/	/	/	0,13	6		[24]
	12 h	2	4	6	8	10	/	6	9		[25]
	24 h	0,065	0,16	0,33	/	/	/	0,16	¼	Solanum lycopersicum	[26]
	4 h	1	3	5	7	/	/	3	12		[27]
	24 h	7,5	/	/	/	/	/	7,5	4	Sesamum indicum	[28]
	4 h	0,0001	0,00015	0,0002	0,00025	0,0003	/	0,00015 0,0003	6	Phaseolus Vulgaris	[29]
	4 h	1,3	1,95	2,6	6,5	/	/	1,95	1,5		[30]
	12 h	5	10	15	/	/	/	5	12	Vigna	[31]

										mungo	
	12 h	0,07	0,13	0,20	0,26	0,33	/	0,20	2	Eruca sativa	[32]
	20 h	0,065	0,130	0,195	0,260	0,325	/	0,325	4	Oryza sativa	[33]
	6 h	0,00001	0,00002	0,00003	0,00004	0,00005	/	0,00004	6	Zea mays	[34]
	6 h	0,1	0,2	0,3	/	/	/	0,3	6		[35]
	6 h	0,1	0,3	/	/	/	/	0,3	6	Arachis hypogaea	[36]
Soak in distilled water	9 h	0,1	0,2	0,3	0,4	/	/	0,4	6	Triticum aestivum	[37]
	6 h	0,2	0,4	0,6	/	/	/	0,2	6		[38]
	12 h	0,07	0,13	0,20	0,26	/	/	0,07	2	Hordeum vulgare	[39]
	4 h	0,098	0,2	0,29	0,4	6,5	/	0,4	2		[40]
	15 h	0,07	/	/	/	/	/	0,07	2		[41]
	4 h	0,01	0,013	0,026	0,052	0,104	/	0,013	2	Capsicum annum	[42]
	6 h	0,1	0,2	0,3	0,4	/	/	0,4	6		[43]
	ND	0,4	0,6	/	/	/	/	0,4	6	Glycine max	[44]
	6 h	0,2	0,4	0,6	0,8	/	/	0,4	9	Sphenostylis stenocarpa	[45]
	6 h	3	5	7	/	/	/	7	5	Brassica campestris	[46]
	6 h	2	4	6	8	/	/	8	5	Brassica napus	[47]
	1 h	0,5	1	2	3	/	/	2 0,5	6	Helichrysum bracteatum	[48]
	6 h	0,1	0,2	0,3	0,4	/	/	0,4	9	Vigna unguiculata	[49]
Buffer solution	2 h	0,001	0,002	0,003	0,004	/	/	0,002	1	Abelmoschus esculentus	[50]
	8 h	0,1	0,2	0,3	0,4	0,5	/	0,5	16	Oryza sativa	[51]
	8 min	0,07	0,13	0,26	/	/	/	0,07	2		[52]

	8 min	0,065	0,130	0,260	/	/	/	0,260		*Pisum sativum*	[53]
	8 min	0,1	0,2	0,3	/	/	/	0,3	4	*Spinach oleracea*	[54]
	5 min	0,033	0,07	0,13	/	/	/	0,07	2,5	*Helianthus annus*	[55]
		0,033	0,07	0,13	/	/	/	0,13	0,5		
Mercuric chloride (HgCl2)	10 h	1	4	7	/	/	/	[illegible]	6	*Dianthus caryophyllus*	[56]
	6 h	0,1	0,2	0,3	/	/	/	0,2	6	*Cicer arietinum*	[57]

Hypochlor ite of 1 min0,001 0,002 0,003 0,004 //0,0023 Sesamum indicum[58]

Sodium 30 minutes 5 10 ///101,5Sorghumbicolor[59]

	ND	0,4	0,8	1,2	1,6	2,0	2,4	2,4	18	*Abelmoschus*	[60]
	ND	0,2	0,5	/	/	/	/	0,5	5	*esculentus*	[61]
	ND	0,5	1	1,5	2	/	/	2	18	*Abelmoschus moschatus*	[62]
	ND	0,165	0,33	0,625	0,125	2.5	/	0,33	24		....
	ND	0,2	0,4	0,6	0,8	1	/	0,03	15	*Khaya senegalensis*	[65]
ND	NA	0,001	0,002	0,003	0,004	/	/	0,4 0,00	4	*Citrullus lanatus* and *Moringa*	[66]
	ND	0,16	0,31	1,25	2,5	/	/	0,16	6	*oleifera*	[67]
	ND	0,10	0,16	/	/	/	/	0,10	18	*Vigna unguiculata*	[68]
	ND	0,1	0,2	0,3	0,4	0,5	/	0,3	24	*Oryza sativa*	[69]
	[illegible]			[illegible]	[illegible]	[illegible]					
	ND	0,130	/	/	/	/	/	3,25	2	*Arachis hypogaea*	[71]
	ND	0,07	0,130	0,260	/	/	/	1			[72]
		0,007			/	/	/	0,13	4	*Hordeum vulgare*	[73]

ND: Pre-treatment not defined; **Ref**: References; **C.A**: Active concentrations. For a good comparative analysis, all concentration values are brought back to the same unit (g/L).

3. Effects induced by the application of sodium azide to seeds

When sodium azide is applied to seeds in an experimental setting, the plant is expected to respond in a variety of ways. Because azides help to create genetic diversity in various crops [74, 75], this mutagen produces Depending on the plant concerned, the effects may be morphological, physiological, biochemical, molecular and/or yield-related (**Table 2**).

Sodium azide (sodium azide) is also considered to be a pro-mutagen, as it generates a reactive organic intermediate that is metabolised in vivo into a potent chemical mutagen in the host plant. This intermediate has been identified in barley as L-azido-alanine. It seems that L-azido-alanine itself does not interact directly with DNA, but that mutagenesis is mediated by its action at the level of cellular processes in the host plant involved in the DNA excision-repair mechanism [76, 77, 78]. Depending on the plant and its concentration, this mutagen can act either to amplify or to reduce a response in the mutant plant.

Table 2: The table shows the effects induced by sodium azide.

Effects induced by sodium azide	Induced mutations	Plants	References
	Reduces the size of the plant	Sesamum indicum ; Triticum aestivum ; Brassicacampestris ; Solanum lycopersicum ; Vigna unguiculata.	[27, 28, 37, 46 9, 58, 63]
	Increases the size of the plant	Eruca sativa; Pisum sativum & or Vicia faba ; Sesamum indicum ; Abelmoschus esculentus	[32, 50, 52, 64]
	Reduces stem diameter	Browallia speciosa ; Khaya senegalensis ; Brassica napus ; Solanum lycopersicum ; Vigna unguiculata	[27, 47, 65, 79]
	Increases the diameter of the stem	Helichrysum bracteatum.	[48]
	Reduces the number of leaves	Khayasenegalensis ; Abelmoschus esculentus ;Solanumlycopersicum ; Hibiscus cannabinus.	[27, 60, 65]
Plant morphology			
	Increases the number of leaves	Sesamum indicum; Citrullus lanatus and Moringa oleifera; Zea mays; Helichrysum bracteatum.	[35, 48, 64, 66]
	Reduces theleaf surface	Eruca sativa; Browallia speciosa; Pisum sativum &orViciafaba ;Khaya senegalensis.	[32, 52, 65, 79]
	Increase the leaf surface	Sesamumindicum ; Erucasativa ; Abelmoschusmoschatus ;Lycopersicon esculentun.	[32, 62, 63, 64,73]
	Reduces length and sheet width	Arachis hypogaea	[70]
	Increases the length and width of the sheet	Arachis hypogaea	[70]
	Reduces the length of the radicle	Trigonella foenum-graecum	[23]

	Increases length of the radicle	Trigonella foenum-graecum	[23]
Plant physiology	Reduces germination	Abelmoschus esculentus ; Sesamum indicum ; Hibiscus cannabinus ; Oryza sativa ; Hordeum vulgare ; Vigna unguiculata	[21, 39, 49, 51, 58]
	Reduction in the pollen fertility	Dianthus caryophyllus ; Triticum aestivum ; Brassica campestris ; Vigna unguiculata	[37, 46, 49, 56]
	Reduces the chlorophylls	Sorghum bicolor; Helianthus annus ; Khaya senegalensis ; Catharanthus roseus	[55, 59, 65, 80]
	Increases chlorophyll content	Eruca sativa ; Vigna unguiculata ; Spinach oleracea ; Trigonella foenum-graecum	[24, 32, 54, 67]
	Reduces the carotenoids	Vicia faba	[52]
	Increases the carotenoids	Helianthus annus; Pisum sativum	[52, 55]
	Reduces sugar content	Sorghum bicolor	[59]
	Increases the sugars	Pisum sativum & Vicia faba ; Trigonella foenum-graecum	[24, 52]
	Reduces the amino acids	ND	ND
	Increases amino acid content	Pisum sativum & Vicia faba	[52]
Plant biochemistry from the	Reduces the proteins	Helianthusannus ;Glycine max ; Trigonella foenum-graecum	[25, 44, 55]
	Increases protein content	Helianthus annus ; Pisum sativum & Vicia faba ; Glycine max ; Trigonella foenum-graecum	[24, 25, 44, 52, 55]
	Reduces the proline	ND	ND
	Increases the proline	Trigonella foenum-graecum	[24, 25]
	Reduces phenol content	Trigonella foenum-graecum	[25]
	Increases the phenols	Pisum sativum; Trigonella foenum-graecum	[25, 53]
	Reduces the flavonoids	Trigonella foenum-graecum	[25]
	Increases flavonoid content	Pisumsativum ;Trigonella foenum-graecum	[25, 53]
	Induces non-lethal point mutations in DNA at a very high rate	Cicer arietinum	[74]

	high		
Plant molecules	Induces chromosomal aberrations at a low rates	Triticum aestivum	[81]
	Causes transitions G : C → A: T and A : T→G :C and a transversion A-T → T-A	Hordeum vulgare ; Zea mays ; Oryza sativa	[34, 41, 82, 83]
	Reduces the number of flowers	Dianthuscaryophyllus ;Abelmoschus moschatus	[56, 62]
	Increases the number of flowers	Abelmoschusmoschatus ;Helichrysum bracteatum	[48, 62]
	Reduces the number of fruits	Lycopersicon esculentun	[72]
	Increases the number of fruits	Cicer arietinum; Glycine max; Arachis hypogaea	[36, 44, 57]
Plant yield	Reduces the number of seeds	Sphenostylis stenocarpa	[45]
	Increases the number of seeds	Glycine max; Capsicum annuum	[44, 43]
	Reduces the weight of fruit	Capsicum annuum	[43]
	Increases the weight of fruit	Lycopersiconesculentun ; Solanum lycopersicum ; Arachis hypogaea	[36, 73, 84]
	Reduces seed weight	Abelmoschus esculentus	[21]
	Increases seed weight	Sesamum indicum; Brassica napus; Zea mays; Arachis hypogaea	[35, 36, 47, 64]
	Reduces fruit yield	Sphenostylis stenocarpa	[45]
	Increases fruit yield	ND	ND
	Reduces seed yield	Sphenostylis stenocarpa	[45]
	Increases seed yield	Brassicacampestris ;Abelmoschus esculentus	[21, 46]
	Diminuela dry biomass of the plant	Khayasenegalensis ; Abelmoschus esculentus ; Hibiscus cannabinus	[22, 60, 65]
	Increases the plant's dry biomass	Sesamum indicum	[22, 64]

ND: Not Defined.

The effects of sodium azide have been revealed in several plant species belonging to a multitude of families (**Table 3**).

Table 3: The table shows the grouping of species by family according to sodium azide induction studies.

Families	Plants
Amaranthaceae	Spinach oleracea
Asteraceae	Helichrysum bracteatum ; Helianthus annus
Apocynaceae	Catharanthus roseus
Brassicaceae	Brassica campestris; Brassica napus; Eruca sativa
Caryophyllaceae	Dianthus caryophyllus
Cucurbitaceae	Citrullus lanatus
Fabaceae	Vigna unguiculata ; Vicia faba ; Arachis hypogaea ; Glycine max ; Pisum sativum ; Trigonella foenum-graecum ; Cicer arietinum ; Sphenostylis stenocarpa
Malvaceae	Abelmoschus esculentus; Abelmoschus moschatus; Hibiscus cannabinus
Meliaceae	Khaya senegalensis
Moringaceae	Moringa oleifera
Poaceae	Triticum aestivum; Zea mays; Oryza sativa; Hordeum vulgare; Sorghum bicolor
Pedaliaceae	Sesamum indicum
Solanaceae	Solanum lycopersicum ; Lycopersicon esculentun ; Capsicum annuum ; Browallia speciosa

4. Interest in a new line of research into mutation induction in okra by application of sodium azide

With its enormous nutritional [85, 86, 87] and medicinal [87, 88, 89] potential, okra is a speculation of interest in the fight against malnutrition worldwide. Despite previous work aimed at improving its importance over time by various means, it is still essential to create new interesting characteristics (which should be retained) using sodium azide (**Table 4**).

Table 4: Previous studies on okra and proposal for a new line of research.

Previous studies				Proposals for a new line of research
	Types of parameters assessed	Overall effect observed	References	
Use of the SA : *Pre-treatment: distilled water (06 h) *Concentrations: 1, 2 and 3 g/L (over 1h) *GE: M1 and M2	Vegetative growth and yield	Reduction compared with the control + variety effect	[21]	Use of the SA :* Pre-treatment : distilled water (pdt 06 h) *Concentrations: 2, 4, 6 and 8 g/L (pdt 06h) * GE: M1; M2 1. Assessment ofin families of metabolites: primary (sugars, proteins, etc.); secondary (phenolic compounds, etc.)
Use of AS + Gamma rays : *Pre-treatment: none *Concentrations: 20, 40, 60, 80 and 100 kr) + 0.130 g/L and (20, 40, 60, 80 and 100 kr) +. 0.195 g/L (pdt 1h) * GE: (M1)	Vegetative growth and yield	Reduction compared with the control + variety effect	[90]	
Use of AS or Gamma rays : *Pre-treatment: none *Concentrations: 0.5, 1.0, 1.5 and 2.0 g/L (for 18 h) *GE: (M1)	Vegetative growth and yield + observation of leaf colouration (green or albino plant)	Decrease in vegetative parameters compared with the control + Increase in yield parameters compared with the control	[62]	
Use of the SA : *Pre-treatment: none *Concentrations : 0.4; 0.8; 1.2; 1.6 and 2.0 g/L (for 18 h) *GE: (M1)	Vegetative growth	Decrease in vegetative parameters compared with the control + variety effect	[60]	by dosage of leaf and/or fruit extracts. 2.Identification of compounds
Use of SA : *Pre-treatment: Buffer solution (2 h) *Concentrations: 0.065; 0.130; 0.195 and 2.60 g/L (pdt 1 h) * GE: (M1)	Vegetative growth and yield	Reduction compared to control	[50]	(catechin, epicatechin, etc.) that are specific to à a biochemical family (whose content deter

Use of the SA : *Pre-treatment: none *Concentrations : 0.2 and 0.5 g/L (for 5 h) *GE: (M1)	Vegetative growth; yield; activities of enzymes (catalase, ascorbate peroxidase) and their expression levels.	No reduction in vegetative growth or yield parameters. Increase in enzyme activity and relative gene expression in treated plants compared with the control.	[61]	mined beforehand). by liquid chromatograp hy (HPLC). 3. Assessment of the mineralogical composition (Mg, Ca, K, Iron, P, etc.) of fruit and/or leaves to determine their nutritional value.

GE: Génération Étudiée; **pdt**: pendant; **AS**: Azoture de Sodium.

In this context of induced mutagenesis, there is an urgent need to define an effective new line of research, complementary to previous studies, that can lead us towards more satisfactory results. In addition, this new line of research will have to be associated with a suitable mutation induction protocol using sodium azide, specific to okra **(Figure 1)**.

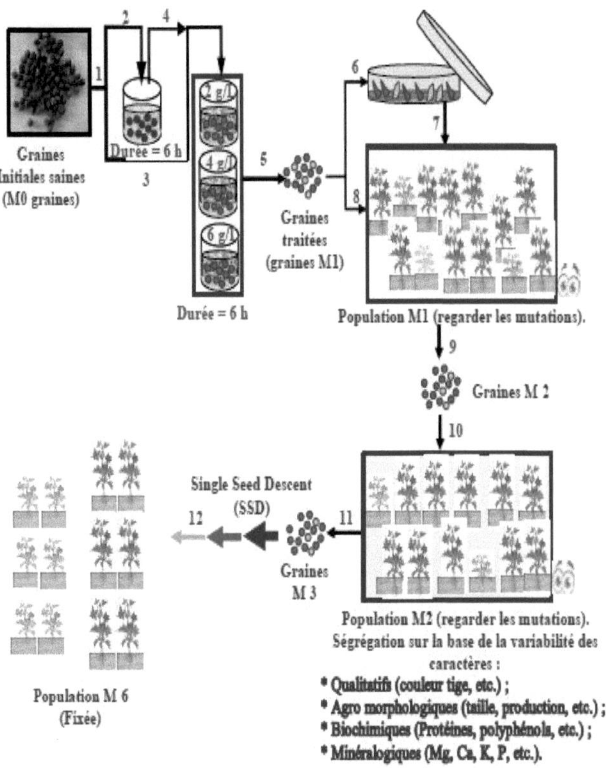

Figure 1: The figure shows a proposed mutation induction procedure for okra using sodium azide (**SA)**. **1**: two possible routes. **2**: seeds are introduced into the distilled water pre-treatment. **3** and **4**: soak the seeds in the SA solutions. **5**: rinsing with water and dewatering of the seeds (to obtain M1 seeds). **6**: germination of the seeds. **7** and **8**: semi-separation (direct for 8) of the seeds in the field to obtain a heterogeneous population (characterisation of the individuals). **9:** production of M2 seeds. **10**: Semi-sowing of M2 seeds to obtain an M2 population (characterisation of its individuals). **11** and **12:** successive self-fertilisation from M2 combined with SSD to obtain homozygous (stable) mutants. This protocol was inspired by those of [91, 92].

5. Agronomic observation of the effect of sodium azide on okra

General observation of plants in the field (90 days after sowing): preliminary trial carried out in **the Bipindi locality**, providing a visual assessment of the development of this crop.

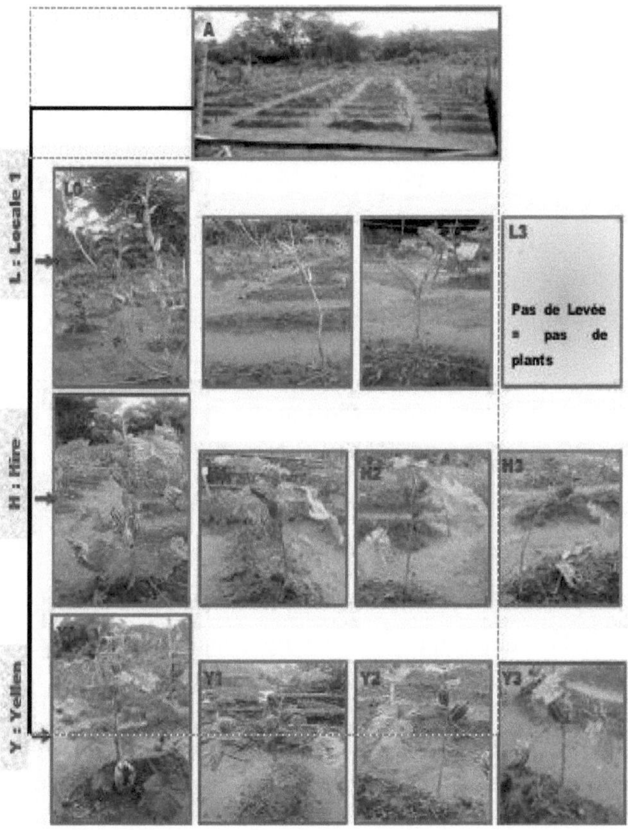

Figure 2: Agronomic responses of okra plants whose seeds have been exposed or not to the mutagen (**NaN3**). **A**: Experimental site; **L0** = **H0** = **Y0** = 0 g/L (**Control**); **L1** = **H1** = **Y1** = 2 g/L (**of NaN3**); **L2** = **H2** = **Y2** = 4 g/L (**of NaN3**); **L3** = **H3** = **Y3** = 6 g/L (**of NaN3**).

6. Conclusion

The aim of this review was to provide an update on the creation of genetic diversity through the application of sodium azide to seeds. While remaining within an experimental framework, success in producing mutants with satisfactory phenotypes by mutagenesis must take into account a number of factors, such as an understanding of the methods of application and the different responses of the plant. The effects of sodium azide are reflected in a range of mutations that can be observed at morphological, physiological, biochemical, molecular and yield levels. Observations made on the effectiveness of sodium azide in its use have highlighted the essential complementarity between pre-treatment, concentration and duration of exposure to this mutagen. In view of this, the water/distilled water pre-treatment generally used favours induction of mutation. This induction on seeds preferably involves the use of concentrations ranging from 1 to 6 g/L, for exposure times of 2 to 6 hours. Standard concentration ranges could therefore be proposed and used for other plants as part of a genetic improvement programme using sodium azide mutagenesis. An effective induction protocol was proposed at the end of this review, taking the case of okra, whose importance and the need to create genetic diversity no longer need to be demonstrated in the fight against food sovereignty in the world. It is therefore essential to carry out a study (practical trial) on the phytochemical characterisation and application of sodium azide (a chemical mutagen) on the seeds of three varieties of okra (Abelmoschus esculentus) grown in Cameroon. The aim of this experiment is to demonstrate the efficiency of this mutagen in improving the nutritional and bioprotective metabolites in this plant.

CHAPTER II

PHYTOCHEMICAL CHARACTERISATION AND APPLICATION OF SODIUM AZIDE (CHEMICAL MUTAGEN) ON THE SEEDS OF

THREE VARIETIES OF OKRA (ABELMOSCHUS ESCULENTUS) GROWN IN CAMEROON.

Raphaël Simeon NJOCK[1*] , Constant Benoit LIKENG-LI-NGUE[1,3] , Jude MANGA NDJAGA[2] , Hermine Bille NGALLE[1] and Joseph Martin BELL[1]

[1] *Université de Yaoundé 1, Faculté des Sciences, Département de Biologie Végétale, Laboratoire de Génétique et Amélioration des Plantes, BP 812 Yaoundé, Cameroon*

[2] *Université de Yaoundé 1, Faculté des Sciences, Département de Biologie Végétale, Laboratoire de Phytopathologie, BP 812 Yaoundé, Cameroon*

[3] *University of Yaoundé 1, Faculty of Science, Department of Plant Biology, Centre de Recherche et d'Accompagnement des Producteurs Agro- pastoraux du Cameroun (CRAPAC), BP 812 Yaoundé, Cameroon*

**Corresponding author: simeonnjocky@gmail.com*

Summary

The aim of this study, carried out at Bipindi in the Southern Region of Cameroon, was to evaluate the phytochemical response of three okra varieties exposed to sodium azide (SA). Healthy seeds were pre-treated with distilled water for 6 h, then exposed to SA at doses of 0, 2, 4 and 6 g/L for 6 h before being sown in the field in a Completely Randomised Block design. Parameters were assessed 60 days after sowing (mature plants, at 50% flowering). The results showed a significant and maximum improvement in values for certain SA concentrations. Sugar content increased by 49.40% (at T2 in M2) and 38.72% (at T5 in M1); protein content increased by 30.13% (at T11 in M1); total flavonoid content increased by 16.27% (at T5 in M2) and 28.91% (at T5 in M1).

% (at T11 in M1). This work shows that applying SA to okra seeds increases the nutritional value (sugars and proteins) and bioprotective value (flavonoids) of this plant. This genetic improvement method could therefore be applied to other plants with high food consumption and medicinal uses that have not yet been tested using this mutagen to improve their metabolite content.

Key words: Sodium azide, okra, phytochemicals, bioprotectants, metabolites.

1. Introduction

Induced mutation is an alternative solution for producing variability in different traits [21], and is applied to the genetic improvement of all attributes, whether qualitative or quantitative [17]. The improved characteristics can result in an increase in both the nutritional [93, 24] and medicinal [25] quality of cultivated plants. Thus, the particular interest in chemical mutagens in the context of induced mutagenesis stems from the fact that they guarantee greater success in selection programmes using vegetative propagation and sexual reproduction [14]. They have also been shown to influence biological changes by DNA base substitution [94], making it possible to induce desirable and exploitable changes in various traits [95]. In view of the number of mutants produced to date, which places it in 8ème position of importance among more than twenty chemical mutagens, the popularisation of the action of sodium azide (AS: NaN_3) on the production of mutants in certain plants such as okra is clearly not observed in the ranking of the speculations preferentially tested [15]. Considered the least dangerous and most effective mutagen, it produces a high number of mutations with moderate sterility rates, associated with few chromosomal aberrations, producing physiological effects capable of inducing a delay in germination and growth [24]. However, an improvement in growth was recorded in certain studies carried out on Abelmoschus moschatus [62], Pisum sativum and Vicia faba [52], Helichrysum bracteatum [48] and Arachis hypogaea [36]. It would therefore be interesting to take a closer look at the effect of this mutagen on okra. Indeed, okra has enormous phytochemical potential to help meet human dietary and medicinal needs. According to some authors, the importance of okra depends on the organ concerned [96]. For example Depending on the part of the plant used, this plant has medicinal properties such as gastroprotective, antioxidant, anti-inflammatory and antimicrobial properties [89]. Thanks to the pectin it contains, it reduces blood cholesterol [87]; its high fibre content regulates blood sugar levels [88]. In addition, okra provides oil rich in unsaturated fatty acids (oleic and linoleic acids) essential for human health [86]. It is also a good source of potassium, sodium, magnesium, calcium, zinc and nickel [85], as well as vitamins B1, B2, B3, B5 and B6 [87]. Today, nearly a billion people are undernourished, particularly in Sub-Saharan Africa (239 million) [97]. Similarly, okra production in this sub-region is far lower than demand in Cameroon alone [98]. By 2080, an additional 266 million people could be at risk of hunger [97]. In addition, given the current level of poverty, more than a billion people still spend more than 10% of their household budget on healthcare [99], so regular consumption of okra, far from being a panacea, could be a great help in alleviating these problems. If the current situation is to be taken into account, it will be necessary to develop okra varieties with an interesting and satisfactory phenotype. A study is therefore needed to assess the variation in metabolites within okra. The aim of this study is to assess the phytochemical response of three okra varieties exposed to sodium azide.

2. Materials and methods

2-1. Presentation of the study site

The trial is being conducted in Cameroon **(Figure 1)**, at the Foyer Notre-Dame de la Forêt (FONDAF) (2°58'8.97" N / 9°56'9.08" E) located in the locality (Arrondissement) of Bipindi, Département de l'Océan, Région du Sud, during the cropping seasons of March - May (M1: first generation) and August - October (M2: second generation) in 2022.

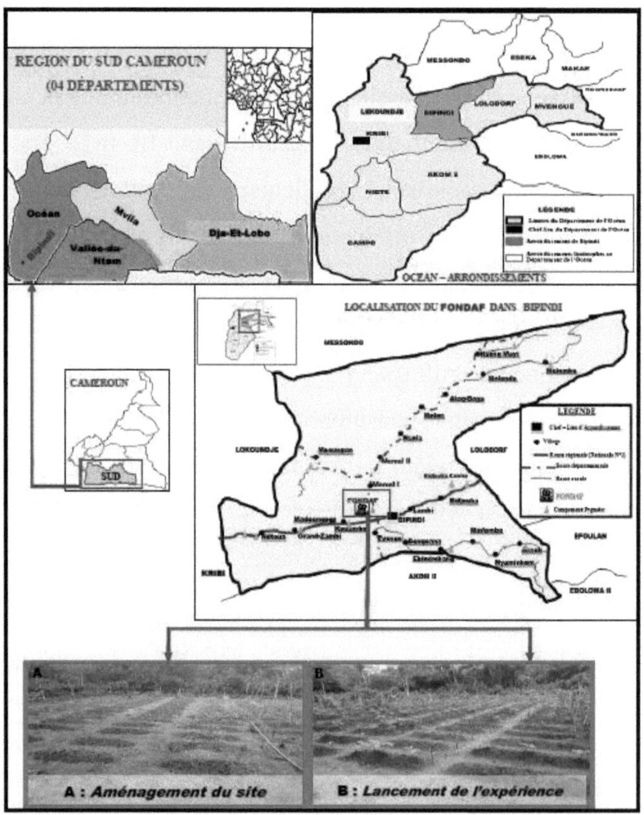

Figure 1: Map showing the location of the study site.

According to its 2015 Communal Development Plan, the Rural Commune of Bipindi, created by Decree No. 95/082 of 24 April 1995, is located 78 kilometres from Kribi, to the south-west of Yaoundé (the political capital). It covers an area of almost 750 km .[2]The climate is humid tropical, equatorial. The average annual rainfall is 1,700 mm and the average temperature is 25.5°C.The relief of this community is part of the vast plateau of southern Cameroon, with an average altitude of 650 m and bordering the coastal plain. It is rugged in places, with isolated hills or hill complexes, varying slopes and the presence of a few rocks. The soils belong to the group of highly desaturated ferralitic soils. They are yellowish-brown to bright brown tropical clay soils. The PH is generally acidic. Very poorly drained hydromorphic soils are also found in the lowlands. The area is watered by a hydrographic network comprising eight main rivers: Lokoundjé, Moungué, Kpwa, Bidjouka, Nsola, Melombo, Tyango, Mougue and their numerous tributaries. These numerous watercourses maintain the climate and supply the homes, including the Foyer Notre Dame de la Forêt (FONDAF), where this experiment was carried out.

2-2. Plant material

The plant material consisted of seeds of three varieties of okra (Abelmoschus esculentus). The Locale 1 variety came from the experimental field of the Unité Génétique Appliquée (UGAP) of the University of Yaoundé 1; the Hire and Yellen varieties were supplied by Agri-Espoir, an approved seed sales company in Cameroon.

2-3. Mutagenic materials

The mutagenic agent used to induce the mutations is sodium azide (AS; NaN_3). It is supplied in Cameroon by the company BST (Biosciences and Technologies).

2-4. Mutation induction protocol

The induction of mutations is carried out according to the protocol below (**Figure 2**). In addition, seeds were sown in the field according to the treatments and the experimental set-up. Data was collected from leaf extracts.

Figure 2: Mutation induction protocol. **A**: Site layout; **B**: Growing plants; **C:** Mature leaves used to collect phytochemical data (**C1:** Locale 1, **C2:** Hire and **C3:** Yellen).

2-5. Experimental set-up

The experimental set-up used is a completely randomised block design, with a set of factors reflecting its characteristics (**Table 1**).

Table 1: Characteristics of the experimental set-up used

Treatments	Meanings	Treatments	Meanings	Treatments	Meanings
T0	Local 1 + 0 g/L sodium azide	T4	Hire + 0 g/L Sodium azide	T8	Yellen + 0 g/L sodium azide
T1	Local 1 + 2 g/L azide of Sodium	T5	Hire + 2 g/L sodium azide Sodium	T9	Yellen + 2 g/L azide of Sodium
T2	Local 1 + 4 g/L azide of Sodium	T6	Hire + 4 g/L sodium azide Sodium	T10	Yellen + 4 g/L azide of Sodium
T3	Local 1 + 6 g/L sodium azide	T7	Hire + 6 g/L Sodium azide	T11	Yellen + 6 g/L sodium azide

2-6. Collection and statistical analysis

The following parameters used to characterise the behaviour of okra plants were assessed 60 days after sowing, i.e. at 50% of plant flowering (mature plants). Six biochemical parameters for each variety were measured using leaf extracts. These were Total chlorophyll (chltotal) [100]; total amino acids (AAT) [101]; total soluble protein (PST) [102]; total soluble sugars (SST) [103]; total polyphenols (PPT) [104]; total flavonoids (FT) [105]. The results for these parameters were subjected to an analysis of variance (ANOVA). Means comparison tests were carried out independently for M1 and M2 using Duncan's method with a threshold of 5% configured in SPSS 18.0 software.

3. Results

3-1. Total Chlorophyll content

Figure 3 illustrates the variation in chlorophyll response as a function of treatment and generation. All three varieties showed similar behaviour, with a significant decrease in the values of this parameter with increasing doses of sodium azide (SA) compared with the control. The decrease in Locale 1 ranged from 20.70 to 23.08% at M1 and from 4.90 to 8.30% at M2, for T1 and T2 respectively. For Hire, it fell from 12.42 to 22.22% in M1 and from 6.04 to 18.36% in M2, from T5 to T7 respectively. In the case of Yellen, it is reduced from 10.18 to 17.49% in M1 and from 2.50 to 9.40% in M2, respectively from Q8 to Q11. These results indicate a more severe action of the mutagen in the first generation. Furthermore, in Locale 1, the dose of NaN_3 applied at T3 would be lethal for this variety.

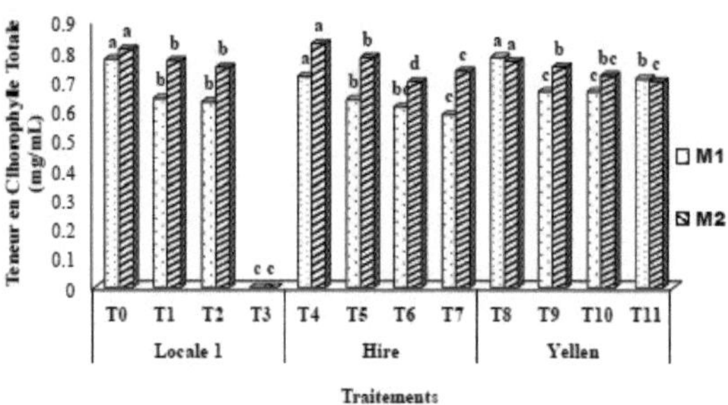

Figure 3: Effect of sodium azide on the Total Chlorophyll content of three okra varieties at M1 and M2.

3-2. Primary metabolite content

Based on **Figure 4**, it can be seen that the sugar response of the three varieties to the treatments used is not similar, reflecting a variety effect. In contrast to the Yellen variety, Locale 1 and Hire recorded a significant increase in their values for the treatments receiving SA. The increase in Locale 1 ranged from 25.30 to 28.90% in M1 and from 25.58 to 49.40% in M2, for T1 and T2 respectively. At Hire, they increased by 38.72% (for T5) in M1 and by 7.50 (for T5) to 20.80% (for T7) in M2. In contrast, Yellen's decline ranges from 36.82% (at T10) to 50.17% (at T11) in M1 and from 4.93% (at T11) to 13.01% (at T10) in M2. Accumulation of this metabolite was greatest at T2 (in M2) and T5 (in M1) for Locale 1 and Hire respectively.

Table 2 shows the quantities of protein metabolites produced by Locale1, Hire and Yellen exposed or not to NaN_3 . The general observation once again reveals a variety effect, which is reflected in the different behaviour of each of the three varieties with increasing doses of NaN .3 With regard to AAT, no significant variation was recorded in Hire at M1 and M2. However, in Yellen there was a significant increase in this metabolite, ranging from 18.70 to 30.05% in M1 from T9 to T11 and 36.25% (for T11) in M2. On the other hand, in Locale 1, values for this parameter fell by 19.30% in M1 and 22.22% in M2 for T2 respectively. In the case of TSP, Yellen showed a significant improvement from 4.20 to 30.13% in M1 and from 5.10 to 28.26% in M2 from T9 to T11 respectively. On the other hand, Hire and Locale 1 saw a significant fall in this parameter. A decrease of 13.87% in M1 and 8.60% in M2 was noted for T7; in addition, a reduction of 3.70 to 4.90% in M1 and 8.04 to 9.87% was noted for T1 and T2 respectively. Accumulation of these metabolites was greater in Yellen, particularly for T11 at M1 and M2.

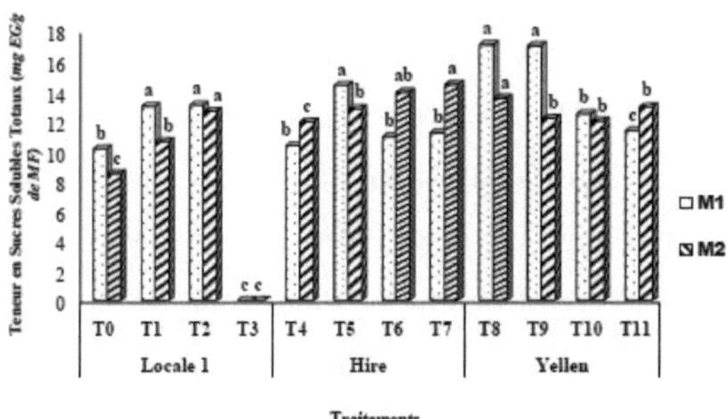

Figure 4: Effect of sodium azide on the Total Soluble Sugar content of the three okra varieties at M1 and M2.

Table 2: Effect of sodium azide on the total amino acid and total soluble protein content of three okra varieties at M1 and M2.

Variétés	Traitements	AAT		PST	
		M1	M2	M1	M2
Locale 1	T0	11,43 a	14,73 a	31,74 a	33,05 a
	T1	10,76 ab	14,31 a	30,24 b	31,16 b
	T2	9,58 b	12,05 b	30,59 b	30,08 b
	T3	0,00 c	0,00 c	0,00 c	0,00 c
Hire	T4	9,26 a	8,34 a	31,03 a	31,67 a
	T5	8,21 a	8,14 a	30,81 a	31,93 a
	T6	8,19 a	7,51 a	31,12 a	31,49 a
	T7	8,35 a	7,85 a	27,25 b	29,16 b
Yellen	T8	8,65 c	7,53 b	25,22 d	24,87 c
	T9	10,27 b	7,48 b	26,28 bc	26,10 b
	T10	10,74 ab	8,20 b	27,95 b	27,45 b
	T11	11,25 a	10,26 a	32,82 a	31,90 a

* **AAT (mg Egly/g MF)** = Total Amino Acids; PST **(mg Eq BSA/g MF) :** Total Soluble Protein.

3-3. Secondary metabolite content

The results of analyses of the effect of sodium azide on the variation in phenolic compounds in Locale 1, Hire and Yellen (**Table 3**) show that, overall, these three varieties show different responses depending on the doses of SA. PPT values showed no significant variations at M1 and M2 for Hire. However, at Yellen, there was a significant increase in the levels of this family of metabolites, varying from 33.47 to 43.24% respectively from T9 to T11 at M1 and from 12.79% (to T11) at M2. On the other hand, in Locale 1 this parameter decreased from 17.48 to 19.08% in M1 and 12.72 to 49.14% in M2 for T1 and T2 respectively.In the case of FTs, Hire and Yellen recorded a significant increase in their values. For Hire, they range from 13.99% in M1 and 16.27% in M2 for T5 respectively. For Yellen, they range from 17.11% to 28.91% in M1 and 4.28% to 22.82% in M2 respectively from T9 to T11. In contrast, Locale 1 values fell significantly from 20.16% to 23.35% in M1 and from 12.82% to 28.57% in M2 for Q1 and Q2 respectively.Accumulation of these metabolites was highest at T11 (at M1) for Yellen in the case of TPP. In the case of FT, it was better at T5 (in M2) and T11 (in M1) for Hire and Yellen respectively.

Table 3: Effects of sodium azide on the total polyphenol and flavonoid content of three okra varieties at M1 and M2.

Variétés	Traitements	PPT		FT	
		M1	M2	M1	M2
Locale 1	T0	26,14 a	27,92 a	16,27 a	17.69 a
	T1	22,25 b	24,77 b	13,54 b	15.68 b
	T2	21,91 b	18,72 c	13,20 b	13.87 c
	T3	0,00 c	0,00 d	0.00 c	0,00 d
Hire	T4	22,25 a	23,09 a	14.34 b	14.87 b
	T5	23,22 a	23,19 a	16,26 a	17,29 a
	T6	22,83 a	22,66 a	14.13 b	14.81 b
	T7	23,42 a	23,01 a	14,62 b	14,97 b
Yellen	T8	16,37 b	22,12 b	11.04 c	13.54 b
	T9	21,85 a	22,14 b	13,32 b	14,12 b
	T10	22,86 a	23,94 ab	14,93 a	15,97 a
	T11	23,45 a	24,95 a	15,53 a	16,63 a

* **PPT (mg EAG/g MF)** = Total Polyphenols; **FT (mg EQ/g MF)** = Total Flavonoids.

4. Discussion

4-1. Total Chlorophyll content

The significant reduction in Chltotal levels in the treatments treated with AS compared with their corresponding controls could be justified by the fact that there was a reduction in the leaf formation process and in the development of the leaf surface, implying a reduction in the production of chlorophyll pigments. A reduction in the values obtained following application of this mutagen to plants such as Sorghum bicolor, Helianthus annuus and Catharanthus roseus has also been noted [59, 55, 80]. Moreover, high doses of this mutagen can stop the enzymes needed for leaf initiation, thus affecting the rate of photosynthesis [43]. In addition, treatment with NaN_3 increases the production of oxidative molecules [53]. The increase in these oxidative molecules is associated with deterioration or destruction of the chloroplasts, including inhibition of chlorophyll synthesis and/or destruction of chlorophyll [55] by peroxidation. [106]. Furthermore, as nitrogen is a fundamental structural constituent of chlorophyll molecules, inhibition of its assimilation due to the action of this mutagen could be the cause of the drop in this pigment. In addition, it has been observed that a deficiency in this mineral disrupts the production and accumulation of chlorophyll in duckweed [107].

4-2. Primary metabolite content

The results obtained, illustrating the significant increase in SST in the Locale 1 and Hire varieties, are linked to a physiological dysfunction induced by this mutagen. This dysfunction would result in a high conversion of atmospheric CO_2 into sugars, followed by a reduction in the consumption of these sugars (in the case of glucose) for cellular metabolism purposes. In fact, its carbohydrates constitute the basic substrate produced by photosynthesis, from which energy (necessary for cell function), amino acids (basic unit of proteins: from carbohydrate carbon chains + nitrogenous minerals from the soil) and proteins (from the cell membrane and enzymes) are produced. The results recorded following the application of this mutagen to Pisum sativum and Vicia faba showed an increase in SST with increasing SA concentration [52]. Thus, the increase in SST content is a defence mechanism of the plant linked to its exposure to the mutagen. The mutagen induces activation of certain genes that are expressed in the direction of accumulation of these osmolytes [108]. The response recorded for AAT and PST reveals an increase in these metabolites in Yellen. The observation for AATs could be explained by an increase in their production, which depends on the combination of amine groups from mineral ammonia and the carbon chains of previously synthesised sugars. The evolution of AAT is also thought to be linked to depolymerisation of the structure of nucleoproteins already formed into amino acids [109]. As for TSP, their increase is thought to be the expression of a regulatory mechanism, in response to the destructuring of essential cellular proteins caused by this mutagen. In fact, to compensate for the plant's need for these metabolites, an acceleration of the transcription and translation processes would be developed, involving a significant synthesis of protein molecules. The results recorded following the application of this mutagen to Pisum sativum and Vicia faba showed an increase in AAT and PST with increasing SA concentration [52]. Furthermore, it was noted that SA treatments induced an increase in PST values compared with controls in Trigonella foenum-graecum [24]. However, the decrease in protein content (observed in Locale 1 and Hire) may be caused by a process of severe protein degradation resulting from increased protease activity [110]. It is also likely that sodium azide caused protein fragmentation due to the toxic effects of reactive oxygen species, which it enables to be produced in vivo and which lead to a reduction in protein content [55]. In addition, it has been observed that the effect of NaN_3 reduces the cellular level of calmodulin, a calcium-binding signal transduction protein involved in the process of cell division [23, 52].

4-3. Secondary metabolite content

The results of analyses revealing an improvement in PPT in Yellen and FT in Hire and Yellen effectively reflect the impact of NaN_3 on the creation of variability within this plant. PPT and FT are molecules produced with medicinal [111] and bioprotective [112] properties against external aggression. It could be that the stressful action of the AS salt has increased the production of these metabolites in these varieties, thereby improving their defence mechanisms and their therapeutic value. Also, based on the results obtained with Hire and Yellen, it was noted that a Another experiment showed that in 80% of cases, an increase in SA concentration caused a change in PPT and FT content in Pisum sativum, Vicia faba and Trigonella foenum-graecum [25, 53]. Flavonoids are frequently induced by abiotic stresses to promote plant protection [113]. They also function as antioxidants that protect membrane lipids from oxidation [114]. Thus, the high levels of phenols and flavonoids found in Trigonella foenum- graecum imply greater (medicinal) antioxidant activity [115]. The latter combats age-related degeneration of cellular components [116].

However, it has been observed that these phenols possess antimutagenic properties that could be involved in mutagen deactivation [117]. They could therefore inactivate the reactive intermediates formed from SA [118]. For example, these metabolites are capable of inhibiting 82% of the mutagenic activity of sodium azide in Salmonella typhimurium [119]. Specifically, they block the transfer of the mutagen into the cytosol by attaching it to transporters on the cell's outer membrane [120]. Phenolic compounds can therefore interact directly and non-enzymatically with the mutagen or form a complex between themselves and the mutagen (NaN_3 - Phenols), thereby reducing the bioavailability of the mutagen [121].In this way, for Hire and Yellen, the increase in their production of these biological compounds following exposure to AI would reflect the setting up by these varieties of a self-regulatory mechanism with the intention of returning or maintaining their metabolism at normal levels. Phenolic compounds are therefore good indicators of the intensity of the action ofa mutagen such as AI in a plant like okra.

5. Conclusion

This work, which evaluated the phytochemical behaviour of three okra varieties exposed to sodium azide, revealed a generally varied response in terms of metabolite production between Locale1, Hire and Yellen. However, there was an improvement in the biochemical response in some of these plants, depending on the treatments. For example, this mutagen induced a significant maximum increase in: sugars at T2 (in M2) for Locale1 and T5 (in M1) for Hire; proteins at T11 (in M1) for Yellen; flavonoids at T5 (in M2) for Hire and T11 (in M1) for Yellen. Mutagenesis by applying SA to okra seeds increases the plant's nutritional value (sugars and proteins) as well as its bioprotective and medicinal value (flavonoids). This genetic improvement method should therefore be extended to other varieties and plants with high food consumption and therapeutic uses not yet tested by this mutagen, to improve their metabolite content.

REFERENCES

[1] - FAO, Sustainably increasing crop production: insights from biological processes. Ed. FAO: Rome, (2011) 1-40.

[2] - FAO, The State of Food and Agriculture. Ed. FAO: Rome, (2016) 1-214.

[3] - FAO, Appropriate seed varieties for small-scale farmers: key practices for DRR practitioners. FAO Ed: Rome, (2014) 44 pages.

[4] - R. A. Laskar and S. Khan, Enhancement of genetic variability through chemical mutagenesis in broad bean. Agro. Technol (2nd International conference on Agricultural and Horticultural Sciences), No. 2 (2014) 1-4 pages.

[5] - O. Yusuff, A. Norhani, Y. R. Ghazali Mohd, R. Asfaliza., A. R. Harun, M. Gous, and U., Magaji, Principle and application of plant mutagenesis in crop improvement: a review. Biotechnology & Biotechnological Equipment, 30(1) (2016) 1-16.

[6] - M. C. A. Kharkwal, brief history of plant mutagenesis. In: Shu Q. Y., Forster B. P., Nakagawa H., editors. Plant mutation breeding and biotechnology (Chap. 24). Wallingford: CABI, (2012) 21 - 30 p.

[7] - S. Diriba, Effects of Colchicine and its Application in Cowpea Improvements
Review Paper. International Journal of Current Innovation Research, 3(9) (2017) 800 - 804.

[8] - J. Bertam, "The molecular biology of cancer". Mol. Aspects Med, 21 (6) (2000) 167-223.

[9] - R. Pathirana, Plant mutation breeding in agriculture: Review. CAB Reviews: Perspectives in Agriculture, veterinary Science, Nutrition and Nutural Ressources, 6(32) (2011) 1-20.

[10] - D. Arulbalachandran, L. Mullainathan and S. Velu, Screening of mutants in black gram (Vigna mungo (L.) Hepper) with effect of DES and COH in M generation. J. Phytol, 1(4) (2009) 213-218.

[11] - R. Roychowdhury & J. Tah, Mutagenesis a potential approach for crop improvement. In: Hakeem K. R., Ahmad P., Ozturk M., editors. Crop improvement: new approaches and modern techniques (Chap. 4). New York (NY): Springer, (2013) 149-187 p.
[12] -. S. Predieri, Mutation induction and tissue culture in improving fruits. Plant Cell Tissue and Organ Culture, 64(2) (2001) 185-210.
[13] -. K. P. M. Dhanayanthi and V. Reddy, Cytogenetic effect of gamma rays and ethyl methane sulphonate in chilli piper (Capsicum annum). Cytologia, 65 (2000) 129-133.
[14] - T. A. Bhat, A. H. Khan and S. Parveen, Comparative analysis of meiotic abnormalities induced by gamma rays, EMS and MMS in Vicia faba L. J. Ind. Bot. Soc. 84 (2005) 45-48.
[15] - Anonymous. Manual of plant improvement by mutation. 3ème edition. FAO/IAEA: Vienna, (2020) 1-270 pages.
[16] - I. Ragunathan and N. Panneerselvam, Antimutagenic potential of curcumin on 16 chromosomal aberrations in Allium cepa. Journal of Zhejiang University Science, 8 (7) (2007) 470-475.
[17] - A. Kleinhofs, W. Owais and R. A. Nilan, Sodium Azide induced mutagenesis in wheat plant. Mut. Res. 55 (1978) 165-195.
[18] - Y. W. Micheale, T. B. Yemane, B. S. Desta, G. A. Girmay, M. J. Hagos, B. K. Abraha, B. S. Fiseha, M. G. Hailay, S. G. Tesfakiros, M. M. Mohammed, M. G. Mullubrhan, K. M. Birhanu, T. Medhin, D. B. Birhanu, and A. G. Haftay, Treatment of seeds with sodium azide for quantitative and qualitative capsule traits at M2 generation of Fourteen Ethiopian sesame (Sesamum indicum L.) genotypes. Heliyon CellPress, 9 (2023) 1-11.
[19] - I. T. Dave, O. A. Neil and J. Van Staden, Colchicine-induced Somatic Polyploids from In Vitro-germinated Seeds of South African Wiatsonia Species. Hortscience, 45(9) (2010) 1398-1402.
[20] - L. Meiya, D. Bin, H. Weipeng, P. Jieli, D. Zhishan, and J. Fusheng,

Induction and Characterization of Tetraploids from Seeds of Bletilla striata (Thunb.) Reichb.f. BioMed Research International, (2018) 1-8.
[21] - N. Mir Muhammad, R. Muhammad, N. Nazo, H. A. N. Syed, H. K. Arif and G. Juma, Effect of chemical mutagens on growth of Okra (Abelmoschus esculentus L. Moench). Pure and Applied Biology, 9(1) (2020) 1110-1117.
[22] - O. I. Ehoniyotan and E. O. Aiyenigba, Effect of sodium azide on agro-morphological traits of four varieties of Kenaf (Hibiscus cannabinus). GSC Biological and Pharmaceutical Sciences, 08(03) (2019) 010-016.
[23] - S. Siddiqui, M. K. Meghvansi, and Z. Hasan, Cytogenetic changes induced by sodium azide (Sodium azide NaN_3) on Trigonella foenum-graecum L. seeds. South African Journal of Botany, 73 (2007) 632-635.
[24] - P. G. Bansod, S. R. Shrivastav and V. A. Athawale, Assessment of Physical and Chemical Mutagenic Effects of Sodium Azide on M1 Generation of Trigonella foenum-graecum L. International Journal of Recent Scientific Research, 10(7) (2019) 33695-33699.
[25] - N. Neha, C. Sana, S. Nidhi, H. Nazarul, A. Al-S. Najla, and A. El-M. Diaa, Frequency and spectrum of M2 mutants and genetic variability in cyto-agronomic characteristics of fenugreek induced by caffeine and sodium azide. Fontiers in Plant Science, (2023) 1-21.
[26] - C. O. M. Sunday, M. O. Regland, O. Opeyemi, S. E. Ifeoma, O. Abisola, and A. A. Muinat, Assessment of genetic divergence in mutant lines of tomato (Solanum lycopersicum L.). International Journal of Engineering Applied Sciences and Technology, 4(7) (2019) 204-210.
[27] - C. A. Adeosun, K. A. Elem, and C. D. Eze, Mutagenic effects of sodium azide on the survival and morphological characters of tomato varieties. Nig. J. Biotech, 37(1) (2020) 55-62.
[28] - Y. W. Micheale, T. B. Yemane, B. S. Desta, G. A. Girmay, M. J. Hagos, B. K. Abraha, B. S. Fiseha, M. G. Hailay, S. G. Tesfakiros, M. M. Mohammed,M. G. Mullubrhan, K. M. Birhanu, T. Medhin, D. B. Birhanu, and

A. G. Haftay, Effect of Sodium Azide on Quantitative and Qualitative Stem Traits in the M2 Generation of Ethiopian Sesame (Sesamum indicum L.) Genotypes. The Scientific World Journal Hindawi, (2021) 1-13.

[29] - S. Abdullahi, B. Y. Abubakar, M. A. Adelanwa, S. A. Shehu, I. M. Zangoma, and A. M. Amshi, Effects of Gamma Rays and Sodium Azide on Yeild Parameters of Phaseolus Vulgaris L. International Journal of Research, 2(11) (2015) 2348-6848.

[30] - M. L. Sorishima, J. O. Joshua, H. K.-N. Emmanuel, T. A. Kpadoo, I. Dennis and E. T. Jeff, Mutagenic Action of Sodium Azide on Germination and Emergence in Landraces of Phaseolus vulgaris L. on the Jos Plateau Agro-Ecological Zone. Journal of Agriculture and Veterinary Science, 10(2) (2017) 64-70.

[31] - A. Asad, H. Salma and U. Kr. Deka, Mutagenic Sensitivity in Early Generation in Black Gram (Vigna mungo L. Hepper). Journal of Agriculture and Veterinary Science, 7(8) (2014) 2319-2372.

[32] - A.-Q. Fahad, Effects of Sodium Azide on Growth and Yield Traits of Eruca sativa (L.). World Applied Sciences Journal, 7 (2) (2009) 220-226.

[33] - G. Vinithashri, S. Manonmani, G. Anand, S. Meena, K. Bhuvaneswari and J. A. John, Mutagenic efficacy and efficiency of sodium azide in rice varieties. Electronic Journal of Plant Breeding, 11(1) (2020) 197-203.

[34] - J. J. Eze and A. Dambo, Mutagenic Effects of Sodium Azide on the Quality of Maize Seeds. Journal of Advanced Laboratory Research in Biology, 6(3) (2015) 76-82.

[35] - B. D. Mustapha, Y. Adamu, I. A. Aminu and A. A. Zainab, Mutagenic effects of sodium azide on some qualitative traits of maize (Zea mays). African Journal of Biological Sciences, 3(1) (2021) 52-57.

[36] - T. A. Eman, Assessment Effect of Mutations on Genetic Variability of Yield and Its Components in some Genotypes of Peanut (Arachis hypogaea L.). J. of Plant Production, 13 (12) (2022) 919-927.

[37] - S. A. Sheikh, M. R. Wani, M. A. Lone, M. A. Tak, and N. A. Malla, Sodium Azide Induced Biological Damage and Variability for Quantitative Traits and Protein Content in Wheat [Triticum aestivum L.]. Journal of Plant Genomics, 2(1) (2012) 34-38.

[38] - P. Srivastava, S. Marker, P. Pandey and D. K. Tiwari, Mutagenic effects of sodium azide on the growth and yield characteristics in wheat (Triticum aestivum L., Em. Thell.). Asian Journal of Plant, 10(3) (2011) 190 - 201.

[39] - B. Dyulgerova and N. Dyulgerov, Mutagenic effect of sodium azide on winter barley cultivars. Agricultural Science and Technology, 14(2) (2022) 27-33.

[40] - A. A. Reem and M. A.-I. Samir, "Effect of Different Concentrations of Sodium Azide on Some traits of Barley (Hordeum Vulgare L.). Acta Scientific Agriculture, 6(8) (2022) 04-08.

[41] - O. Olsen, W. Xingzhi and V. W. Diter, Sodium azide mutagenesis : Preferential generation of AT -> GC transitions in the barley Antl8 gene Proc. Natl. Acad. Sci. 90 (1993) 8043-8047.

[42] - Yafizham, and B. Herwibawa, The effects of sodium azide on seed germination and seedling growth of chili pepper (Capsicum annum L. cv. Landung). International Symposium on Food and Agro-biodiversity, 102 (2017) 1-5.

[43] - J. O. Omeke, E. O. Ojua, N. M. Eze, and N. E. Abu, Effect of sodium azide induction on morphological traits of shombo and tatase (Capsicum annuum L.). Nigerian Agricultural Journal, 52(1) (2021) 111-117.

[44] - H. Sajid, A. Zahid, S. Ghulam and A. Qadeer, Response of Soybean Genotypes to Different Levels of Mutagens for Yield Related Attributes. Biomed J. Sci. & Tech. Res. 25(1) (2020) 18757-18764.

[45] - T. A. Solomon, A. A. Johnson, and C. U. Paul, Impact of sodium azide on maturity, seed and tuber yields in M1 African yam bean [Sphenostylis stenocarpa (Hochst ex A. Rich) Harms] generation. Trop. Agric. 98 (1) (2021)

16-27.
[46] - G. Kumar & K. Dwivedi, Sodium Azide Induced Complementary Effect of Chromosomal Stickiness in Brassica campestris L. Jordan Journal of Biological Sciences, 6(2) (2013) 85-90.
[47] - H. Saddam, M. K. Wisal, S. K. Muhammad, A. Naveed, U. Nosheen, A. Sajjad, A. S. S. Sajjad, and Syed Mutagenic effect of sodium azide (NaN_3) on M2 generation of Brassica napus L. (variety Dunkled). Pure and Applied Biology, 6(1) (2017) 226-236.
[48] - M. A. El-Khateeb, A. Rawia, K. H. I. Eid Hashish, H. A. Ashour and R. M. S. Radwan, Determination of the effective of chemical mutagenesis using sodium azide to improvement of vegetative growth and flowering characteristics in Helichrysum bracteatum L. Plant Journal of Pharmaceutical Negative Results, 13(3) (2022) 1894-1904.
[49] - R. Aamir, R. T. Younas and K. Samiullah, Assessment of Bio physiological damages and cytological aberrations in cowpea varieties treated with gamma rays and sodium azide. PLoS ONE, 18(7) (2022) 1-26.
[50] - H. M. Ati, Effect of Gamma Ray and Sodium Azide on the Germination, Survival and Morphology of Varieties of Okro (Abelmoschus esculentus L. moench). International Journal of Horticulture & Agriculture, 2(1) (2017) 1-4.
[51] - S. Kirti, K. S. Alok, D. K. Dwivedi, N. A. Khan and S. P. Singh, Effects of sodium azide on seed germination and seedling growth in Kalanamak rice. The Pharma Innovation Journal, 11(7) (2022) 711-713.
[52] - K. M. Saad-Allah, M. Hammouda and W. A. Kasim, Effect of sodium azide on growth criteria, some metabolites, mitotic index and chromosomal abnormalities in Pisum sativum and Vicia faba. International Journal of Agronomy and Agricultural Research, 4(4) (2014) 129-147.
[53] - M. Hamouda, K. M. Saad-Allah and W. A. Kasim, Molecular and physiological responses of Pisum sativum and Vicia faba to sodium azide. IJAAR. 4(6) (2014) 46-61.

[54] - R. Srivastava, A. Jagrati, P. Mahima and V. Anugunja, Mutagenic Effect of Sodium Azide (NaN) on Seed Germination and Chlorophyll Content of Spinach oleracea, Ind. J. Pure App. Biosci. 7(4) (2019) 366-370.
[55] - S. Elfeky, S. Abo-Hamad and K. M. Saad-Allah, Physiological impact of sodium azide on Helianthus annus seedlings. International Journal of Agronomy and Agricultural Research, 4(5) (2014) 102-109.
[56] - R. Rajib and T. Jagatpati, Chemical mutagenic action on seed germination and related agro-metrical traits in M1 Dianthus generation. Current Botany, 2(8) (2011) 19-23.
[57] - P. K. Mahesh and K. S. Vijay, Effects of sodium azide on yield parameters of chickpea (Cicer arietinum L.). Journal of Phytology, 3(1) (2011) 39-42.
[58] - K. D. Abubakar and L. S. Abdu, Effects of para-dichlorobenzene and sodium azide on germination and seedling growth of sesame (Sesamum indicum L.). FUDMA Journal of Sciences (FJS), 5(2) (2021) 203-207.
[59] - D. M. Umar, R. Erum and M. Rafiq, Physical and biochemical analysis of sodium azide treated Sorghum bicolor (L.) Monech. Pak. J. Biotechnol, 8(2) (2011) 67-72.
[60] - S. D. Amol, S. D. Asmita, S. P. Sweety, G. S. Nileema and H. Sanjay, Effect of Sodium Azide Induction on Germination Percentage and Morphological Growth in Two Varieties of. Int. J. Curr. Microbiol. App. Sci. 7(6) (2018) 3586-3593.
[61] - D. V. M. A. Emuejevoke, M. E. El-Esawi, A. A. Imoni, H. M. Al-Ghamdi, M. M. Ali, E. El-Sheekh, A. Abdeldaym and Al-D. A. Monerah, Sodium Azide Priming Enhances Waterlogging Stress Tolerance in Okra (Abelmoschus esculentus L.). Agronomy, 9(679) (2019) 1 - 16.
[62] - W. R. Ashish, R. H. Nandkishor and W. Prashant, Effect of sodium azide and gamma rays treatments on percentage germination, survival, morphological variation and chlorophyll mutation in musk okra (Abelmoschus moschatus l.).

Int. J. Pharm. Sci. 3(5) (2011) 483-486.
[63] - J. K. Mensah, B. Obadoni, P. A. Akomeah, B. Ikhajiagbe and J. Ajibolu, The effects of sodium azide and colchicine treatments on morphological and yield traits of sesame seed (Sesame indicum L.). African Journal of Biotechnology, 6(5) (2007) 534-538.
[64] - R. E. Aliyu, A. Aliyu and A. K. Adamu, Induction of Genetic Variability in Sesame (Sesamum indicum L.) with Sodium Azide. FUW Trends in Science & Technology Journal, 2(2) (2017) 964 - 968.
[65] - M. G. Gehan, Effect of some Chemical Mutagens on the Growth, Phytochemical Composition and Induction of Mutations in Khaya senegalensis. Int. J. Plant Breed. Genet. 9 (2) (2015) 57-67.
[66] - A. A. AbdulRahaman, A. A. Afolabi, D. A. Zhigila, F. A. Oladele and A. A. Al Sahli, Morpho-anatomical effects of sodium azide and nitrous acid on Citrullus lanatus (Thunb.) Matsum. & Nakai (Cucurbitaceae) and Moringa oleifera Lam. (Moringaceae). Hoehnea, 45(2) (2018) 225-237.
[67] - B. P. Mshembula, J. K. Mensah and B. Ikhajiagbe, Comparative assessment of the mutagenic effects of sodium azide on some selected growth and yield parameters of five accessions of cowpea - Tvu-3615, Tvu-2521, Tvu-3541, Tvu-3485 and Tvu-3574. Archives of Applied Science Research, 4 (4) (2012) 1682-1691.
[68] - D. Kumala, M. Gita and S. Sudjino, Effects of Sodium Azide (NaN_3) And Cytokininon Vegetative Growth and Yield of Black Rice Plant (Oryza sativa L. 'Cempo Ireng). Advances of Science and Technology for Society, (2016) 130005-1-130005-12.
[69] - J. K. Mensah and B. Obadoni, Effects of sodium azide on yield parameters of groundnut (Arachis hypogaea L.). African Journal of Biotechnology, 6(6) (2007) 668-671.
[70] - D. A. Animasaun, S. Oyedeji, M. A. Azeez and A. O. Onasanya, Evaluation of the Vegetative and Yield Performances of Groundnut (Arachis

hypogaea) Varieties Samnut 10 and Samnut 20 Treated with Sodium Azide. International Journal of Scientific and Research Publications, 4(3) (2014) 1-10.

[71] - B. Dyulgerova and N. Dyulgerov, Grain yield and yield related traits of sodium azide induced barley mutant lines. Journal of Central European Agriculture, 21(1) (2020) 83-91.

[72] - A. K. Adamu & H. Aliyu, Morphogical effects of sodium azide on Tomato (Lycopersicon esculentum Mill). Science World Journal, 2(4) (2007) 9-12.

[73] - Y. Aminu, B. U. Bala, H. I. Kabiru and A. A. Musbahu, Induced growth and yield responses to seasonal variation by sodium azide in tomato (Lycopersicon esculentum Mill.). Bayero Journal of Pure and Applied Sciences, 10(1) (2017) 226-230.

[74] - M. R. Wani, M. I. Kozgar, N. Tomlekova and al. Mutation breeding: a novel technique for genetic improvement of pulse crops particularly Chickpea (Cicer arietinum L.). In: Parvaiz A., Wani M. R., Azooz M. M., Lam-son P. T., editors. Improvement of crops in the era of climatic changes (Chap. 9). New York (NY): Springer, (2014) 217-248 pp.

[75] - R. Roychowdhury and J. Tah, Chemical mutagenic action on seed germination and related agro-metrical traits in M1 Dianthus generation. Current Botany, 2(8) (2011) 19-23.

[76] - W. Owais and A. Kleinhofs, Metabolic activation of the mutagen azide in biological systems. Mut. Res. Fundamental and Molecular Mechanisms of Mutagenesis, 197(2) (1988) 313-323.

[77] - M. F. Sadiq and Z. M. Owais Mutagenicity of sodium azide and its metabolite azidoalanine in Drosophila melanogaster. Mut. Genetic Toxicology and Environmental Mutagenesis, 469(2) (2000) 253-257.

[78] - V. E. Viana, C. Pegoraro, C. Busanello and D. O. A. Costa, Mutagenesis in rice: the basis for breeding a new super plant. Frontiers in Plant Science, 10(1326) (2019) 1-28.

[79] - H. E. El-Mokadem and G. G. Mostafa, Induction of mutations in Browallia speciosa using sodium azide and identification of the genetic variation by peroxidase isozyme. African Journal of Biotechnology, 13(1) (2013) 106-111.

[80] - V. Mistry, T. Pragya, P. Paresh, S. V. Gajendra, L. Geung-Joo and S. Abhishek, Ethyl Methane Sulfonate and Sodium Azide-Mediated Chemical and X-ray-Mediated Physical Mutagenesis Positively Regulate Peroxidase 1 Gene Activity and Biosynthesis of Antineoplastic Vinblastine in Catharanthus roseus. Plants, 11(2885) (2022) 1-27.

[81] - D. Shailja, B. Renu and M. Shrilekha, Sodium azide induced mutagenesis in wheat plant. World Journal of Pharmacy and Pharmaceutical Sciences, 6(10) (2017) 294 - 304.

[82] - V. Talamè, R. Bovina, M. C. Sanguineti, R. Tuberosa, U. Lundqvist and S. Salvi, TILLMore, a resource for the discovery of chemically induced mutants in barley. Plant Biotechnol. J., 6(5) (2008) 477-485.

[83] - T. H. Tai, A. Chun, I. M. Henry, K. J. Ngo and D. Burkart-Waco, Effectiveness of sodium azide alone compared to sodium azide with methyl nitrosurea for rice mutagenesis. Plant Breed. Biotechnol, 4(4) (2016) 453- 461.

[84] - B. U. Bala, S. I. Yelwa, F. S. Hassan, and S. M. Babangida, Mutagenic effects of sodium azide (NaN3) on morphological characteristics on two varieties of tomato (Solanum lycopersicum Mill). Bayero Journal of Pure and Applied Sciences, 11(1) (2018) 50 - 54.

[85] - E. I. Moyin-Jesu, Use of plant residues for improving soil fertility pod nutrients root growth and pod weight of okra Abelmoschus esculentum L. Bioresour, Tech, 98 (2007) 2057-2064.

[86] - Anonymous, Series of crop specific biology documents : Biology of Abelmoschus esculentus L. Ed. Department of Biotechnology: Ministry of Science and Technology Government of India, (2011) 1-35 pages.

[87] - E. C. Legba, R. G. Ahlincou, L. R. Atchanhouin, L. A. Aglinglo, R. A.

Francisco, H. Fassinou, V. Nicodème, and E. G. Achigan-Dako, Fiche technique synthetique pour la production du gombo (Abelmoschus esculentus L.). Publisher, Laboratory of Genetics Horticulture and Seed Science (GBioS). ResearchGate, 106(69) (2021) 1-6 pages.

[88] - T. Ngoc, N. Ngo, T. Van and V. Phung, Hypolipidemic effect of extracts from Abelmoschus esculentus L. (Malvaceae) on Tyloxapol-induced hyperlipidemia in mice. Warasan Phesatchasat, 35(1-4) (2008) 42-46.

[89] - A. Ali and S. S. Deokule, Comparison of phenolic compounds of some edible plant of Iran and India. Pak. J. Nutr. 7(4) (2008) 582-585.

[90] - H. M. Ati and A. K. Adamu, Effect of combined doses of gamma ray and sodium azide (mutagenic agents) on the morphological traits of some varieties of okra (Abelmoschus esculentus). African Journal of Agricultural, 11(32) (2016) 2968-2973.

[91] - M. A. J. Parry, J. M. Pippa, B. Carlos, T. Katie, H.-L. Antonio, B. Marcela, R. Mariann and L. P. Andrew, Mutation discovery for crop improvement. Journal of Experimental Botany, 60(10) (2009) 2817-2825.

[92] - S. Xavier, E. Roger, G. Nathalie, M. Luisa, N. Salvador and L. Eric, EMS mutagenesis in mature seed-derived rice calli as a new method for rapidly obtaining TILLING mutant populations. PLANT METHODS, 10(5) (2014) 1-13.

[93] - G. K. Navnath & P. K. Mukund, Effect of physical and chemical mutagens on pod length in Okra (Abelmoschus esculentus L. Moench). Sci. Res. Reporter, 4(2) (2014) 151-154.

[94] - J. L. Cooper, B. J. Till, Laport R. G., Darlow M. C., Kleffner J. M., Jamai A., Tarik El-M., Shiming L., Rae R., Niels N., Kristin D. B., Khalid M., Luca C. and Steven H., 2008. TILLING to detect induced mutations in soybean. BMC Plant Biol, 8(9): 1-10.

[95] - N. Gupta, S. Sood, Y. Singh & D. Sood, Determination of lethal dose for gamma rays and ethyl methane sulphonate induced mutagenesis in okra

(Abelmoschus esculentus (L.) Moench.). SABRAO J. Breed. Genet, 48 (3) (2016) 344-351.

[96] - Y. Mihretu, G. Wayessa & D. Adugna, Multivariate Analysis among Okra (Abelmoschus esculentus (L.) Moench) Collection in South Western Ethiopia. Journal of Plant Sciences, 9(2) (2014) 43-50.

[97] - S. Capot & D. Perus, Les Enjeux De La Propriété Industrielle Appliques Aux Semences Végétales. Cahiers Du Lab. RII : Documents De Travail, Université de la côte d'opale, Laboratoire de Recherche sur l'Industrie et l'Innovation. N°269 (2013) 1 - 27.

[98] - Anonymous, Growth and Employment Strategy Paper. Publisher: MINEPAT. Yaoundé, Cameroon, (2009) 112p.

[99] - Anonymous, Reorienting health systems towards primary health care as the resilient foundation of universal health coverage, and preparations for a high-level meeting of the United Nations General Assembly on universal health coverage: Report of the Director-General. Publisher: WHO, (2023) 11 p.

[100] - D. I. Arnon, Copper enzymes isolated chloroplast, polyphenoloxidase in Beta vulgaris. Plant physiology, 24 (1949) 1 - 15.

[101] - E. M. Yemm & E. C. Cocking, The determination of amino acids with ninhydrin. The Analyst, 80 (1955) 209-213.

[102] - M. Bradford, A rapid and sensitive method for the quantitation of microgram quantities of protein utilizing the principle of protein-dye binding. Anal. Biochem, 7 (72) (1976) 115-125.

[103] - E. M. Yemm & A. J. Willis, The estimation of carbohydrates in plants extracts by Anthron. Biochemistry Journal, 9(2) (1954) 27-36.

[104] - G. Marigo, Méthode de fractionnement et d'estimation des composes phénoliques chez les végétaux. Analysis, 2 (1973) 106-110.

[105] - O. A. Aiyegoro & A. Okoh, Preliminary phytochemical screening and in vitro antioxidant activities of the aqueous extract of Helichrysum longifolium DC. BMC Complementary & Alternative Medicine, 10 (21) (2010) 1 - 9.

[106] - A. H. Price & G. A. F. Henry, Iron catalase oxygen radical formation and its possible contribution to drought damage in nine native grasses three cereals. Plant Cell and Environment, 14 (1991) 477-484.

[107] - S. S. Demirors, G. Keser & M. Dogan, Effects of lead on chlorophyll content, total nitrogen, and antioxidant enzyme activities in duckweed (Lemna minor). Int. J. Agric. Biol. 15(1) (2013) 145-148.

[108] - N. M. El-Shafey, R. A. Hassaneen, M. A. Gabr & O. ElSheihyd, Pre-exposure to gamma rays alleviates the harmful effect of drought on the embryo-derived rice calli. Australian Journal of Crop Science, 3 (5) (2009) 268-277.

[109] - G. Kumar, S. Kesarwani & V. Sharma, Clastogenic effect of individual and combined treatment of Gamma rays and EMS in Lens culinary. Journal of Cytology and Genetics, 4 (2003) 149-154.

[110] - M. J. Palma, M. L. Sandalio, J. F. Corpas & Romero-P., Plant proteases, protein degradation, and oxidative stress: role of peroxysomes. Plant Journal and Biochemistry. 40(6) (2002) 521-530.

[111] - A. E. O. Elkhalifa, E. Alshammari, M. Adnan, J. C. Alcantara, A. M. Awadelkareem, N. E. Eltoum, K. Mehmood, B. P. Panda & S. A. Ashraf, Okra (Abelmoschus esculentus) as a Potential Dietary Medicine with Nutraceutical Importance for Sustainable Health Applications. Molecules, 26 (2021) 696.

[112] - E. H. D. Fotso, A. C. Djeuani, D. M. Djamndo & N. D. Omokolo, Evaluation of polyphenoloxidase and peroxidase activities and the accumulation of phenolic compounds in the resistance of cassava stimulated with Benzo (1,2,3) thiadiazol-7-carbothionic acid-s-methyl ester against Colletotrichum gloeosporioides Penz. Int. J. Biol. Chem. Sci. 15(3) (2021) 950-965.

[113] - R. A. Dixon and N. L. Paiva, Stress-induced phenylpropanoid metabolism. Plant Cell, 7 (1995) 1085-1097.

[114] - J. E. Li, S. T. Fan, Z. H. Qiu, C. Li & S. P. Nie, Total flavonoids content, antioxidant and antimicrobial activities of extracts from Mosla chinensis maxim. Jiangxiangru. LWT Food Sci. Technol, 64 (2015) 1022-1027.

[115] - H. Noreen, N. Semmar, M. Farman & J. S. O. McCullagh, Measurement of total phenolic content and antioxidant activity of aerial parts of medicinal plant Coronopus didymus. Asian Pac. J. Trop. Med, 10 (2017) 792-801.
[116] - C. T. Sulaiman & I. Balachandran, Total phenolics and total flavonoids in selected Indian medicinal plants. Indian J. Pharm. Sci. 74 (2012) 258-260.
[117] - R. C. Hornl & V. M. F. Vargas, Antimutagenic activity of extracts of natural substances in the Salmonella/microsome assay. Mutagenesis, 18(2) (2003) 113-118.
[118] - S. Gowri & P. Chinnaswamy, Evaluation of in vitro antimutagenic activity of Caralluma adscendens Roxb in bacterial reverse mutation assay. Journal of Natural Products and Plant Resources, 1(4) (2011) 27-34.
[119] - L. Birossov, M. Mikulasov & S. Vaverkov, Antimutagenic effect of phenolic acids. Biomed Pap Med Fac Univ Palacky Olomouc Czech Repub, 149 (2) (2005) 489-491.
[120] - T. Hour, Y. Liang, I. Chu & J. Lin, Inhibition of eleven mutagens by various tea extracts, (-) Epigallocatechin-3-gallate, Gallic Acid and Caffeine. Food and Chemical Toxicology, 37 (1999) 569-579.
[121] - E. G. De Mejia, E. Castano-Tostado & G. Loarca-Pina, Antimutageniceffects of natural phenolic compounds in beans. Mutation Research, 441 (1999) 1-9.

TABLE OF CONTENTS

Printed by Books on Demand GmbH, Norderstedt / Germany